ELEPHANTS

To the wildlife guides who helped
make this book possible.

All rights reserved. Published by Scholastic Press,
an imprint of Scholastic Inc., *Publishers since 1920.*
SCHOLASTIC, SCHOLASTIC PRESS, and associated logos
are trademarks and/or registered trademarks
of Scholastic Inc.

Library of Congress Cataloging-in-Publication
Data available

ISBN 978-0-545-60580-9
10 9 8 7 6 5 4 3 2 1   22 23 24 25 26

Printed in China   62
First edition, November 2022

Cover and title page photo: African savanna elephant and Mount Kilimanjaro.
Photo © 6-7: Mark MacEwen/Nature Picture Library.

Book design by Three Dogs Design LLC.

NIC BISHOP
ELEPHANTS

Scholastic Press • New York

THE MIGHTY ELEPHANTS

Elephants are by far the biggest of all land animals. A large male can toss a lion into the air with its trunk. Some have tusks taller than a person. And yet, elephants are more than mighty. They are sensitive and intelligent creatures, too. They live in large, caring family societies run by females. They have extraordinary senses and communicate in complex ways that scientists are only now starting to understand.

This old African savanna elephant has a broken tusk, which can happen in battle with another elephant. The broken tusk will not be replaced with a new one, but it can continue to grow in length.

ELEPHANT TYPES

There are three types of elephants. The largest is the African savanna elephant, which roams the open forests and grasslands of South and East Africa. An adult male can stand twelve feet tall at the shoulders and weigh as much as 13,000 pounds. The next in size is the Asian elephant, which lives in the forests and grasslands of India and Southeast Asia. A full-grown male stands nine feet tall at the shoulders and weighs around 8,500 pounds. Its ears are smaller than a savanna elephant's and do not reach up over the neck. Also, female Asian elephants do not have visible tusks, nor do some of the males.

Deep in the rain forests of Borneo, there are elephants called Bornean elephants. They are a type of Asian elephant but are a foot or so shorter. Bornean elephants are shy and very hard to find in the dense forest. This one has come to the river for a drink.

This family group of African forest elephants has
entered a clearing to meet other elephants and
feed on minerals they obtain from the mud.

The third type, and perhaps the smallest, is the African forest elephant. An adult male is about eight feet tall at the shoulders and weighs about 8,000 pounds. Forest elephants are secretive, and we still do not know a lot about them. They travel in small family groups through the dense rain forests of West Africa, particularly in the Congo Basin. They have less-wrinkled skin and rounder ears than savanna elephants. They also have straighter tusks that point more downward to avoid becoming tangled with branches and bushes in the undergrowth.

ELEPHANT CALVES

The birth of an elephant is a time of celebration. Other members of the herd gather around, trumpeting with excitement. They are curious and eager to help the new calf. Trunks reach down to tenderly touch the wobbly youngster and help it to its feet.

An elephant herd provides a caring community for a calf. It is a family group made up of females and their calves. Among African savanna elephants and Asian elephants, the herd usually has six to twenty members. It often includes grandmothers, aunties, elder sisters, and their calves. African forest elephants, however, live in smaller groups. The herd may be just the mother and her calves.

These Asian elephants are sheltering a young calf within the security of their trunks and legs.

An elephant mother is always patient with her calf. It will depend on her for about four years, or until it is weaned. The calf rarely strays far from its mother, and if it does, an elder sister or auntie will be watching to make sure it doesn't get into trouble. These helpers are called allomothers, because they are almost like second mothers. They will rescue the calf if it falls into a water hole, and protect it if it gets bullied. An allomother will often comfort a calf when it is upset by soothing it with a lullaby of soft rumbling calls.

Elephant calves suck their trunks, in much the same way that human infants suck their thumbs. The white bird is an egret, which eats insects disturbed by the elephants as they walk around.

HELPING THE HERD

Elephants live for about sixty years and have excellent memories. That means that an old female will have a lot of knowledge and wisdom she can use to help her herd. This senior female has an especially important role among African savanna elephants. She is called a matriarch and acts as a herd leader. During a drought she will remember where the herd went last time to find water. There are stories of African savanna elephants marching across desert

sand dunes to find a special water hole they had not visited in over twenty years. The matriarch will know which trails to follow, where to find food at different times of the year, and how to safely cross a river. She will recognize dangerous predators and know when to call the herd into a defensive huddle, guarding the youngest behind a forest of legs. The matriarch will also know a large network of friends and allies among other elephant herds. A single elephant is thought to be able to recognize about a hundred other elephants by sound, smell, and touch. They can even recognize old friends more than twenty years after last meeting them.

These African savanna elephants are crossing a river as a group, led by the matriarch. You can see that the youngest has had to swim.

An elephant herd will often meet up with neighboring herds.

These are family groups that are related to one another. Elephants are very sociable, so these meetings are noisy, with a happy chorus of trumpeting and rumbling calls as old friends gather and greet one another.

The group may travel together to a water hole, enjoying the safety of numbers. African savanna elephants may be led by the senior matriarch, and everyone will keep order. The older elephants drink at the best spots, and nobody splashes or muddies the water till everyone has had their fill. An elephant drinks about fifty gallons of water a day.

During times of drought, African savanna elephants may have to go without drinking for three or four days until the matriarch has led them to a reliable source of water. These elephants have just reached the Chobe River in Africa and are eager to quench their thirst.

PLAYTIME AND COMMUNICATION

For calves, these gatherings are a chance to find new playmates and make friendships that last a lifetime.

Calves love to play, especially in their first year. A friendly head waggle is an invitation to wrestle, chase, or run in circles. Sometimes calves take turns playing with a toy, such as a large leaf or a piece of paper.

Young calves like to play tag and engage in mock battles.

By itself, a calf may seem to play with an imaginary friend, chasing it and then fleeing in wide-eyed excitement.

Elephants are very expressive with their bodies. It is a way they communicate with each other. For example, an angry elephant will stand tall and flare out its ears. If impatient, it will shake its head and ears, throwing off clouds of dust. A showing-off elephant will walk with a jaunty, bouncing swagger. One that feels awkward will keep touching its face with its trunk, as if the elephant is trying to reassure itself.

SOUND THE TRUMPET

But elephants communicate mostly with sound. One call they make is the trumpet, which is made by blowing hard down the trunk. A trumpet call is like an exclamation mark. It is made when an elephant is surprised, excited, or angry. Other calls are made by blowing across the vocal cords, in the same way that we make sounds. Scientists have given names to the different varieties of these sounds, such as barks, grunts, and revs. Some of these sounds, called roars, are loud and used when an elephant is alarmed or wants to scare predators. Most elephant calls, though, are called rumbles. These have a purring-growling sound.

When elephant groups meet, they greet each other with soft rumbling calls. They also use their trunks to gently touch and smell each other. It is their way of checking each other out and saying "hello."

ELEPHANT "SPEAK"

Elephants make special greeting rumbles
when they meet. There are other rumbles
meaning "I'm okay," "I'm here," and "I'm worried."
A calf has calls that it makes when it is lost, hungry, or
being bullied by an older calf. In fact, scientists have
identified dozens of different calls, and the more

they study these complicated sounds, the more

they are finding that elephants have a type of

language, and can even discuss things. For example,

when an African savanna elephant matriarch decides

it is time to leave a water hole, she points her body in

the way she wants the herd to go and makes a "let's

go" rumble. But sometimes other members of the

herd disagree and there is a rumbling conversation.

It's as if they are discussing different options, until a

decision is made on where to go.

This herd has gathered around a water hole in Botswana to drink. A group will rarely stay long, because there may be other elephant herds waiting for a turn, or perhaps predators nearby. The elephant with her trunk raised is probably the matriarch checking for the scent of other arriving elephants.

Elephants also communicate by smell.
They produce chemical signals in their urine and dung, as well as from glands called the temporal glands that are located between the eyes and ears. Elephants produce scents when they are excited, stressed, or looking for a mate. Other elephants can detect this, even from miles away, because they have an extraordinary sense of smell. Scientists believe elephants have a better sense of smell than bloodhounds.

This sense of smell is made more special thanks to an elephant's trunk. It is a wonderful mobile sniffer. An elephant can wave its trunk in the air to pick up distant smells, or poke its trunk into a crevice for a spot check. And to learn more detail, it can rub its trunk on a surface and then wipe it on a special organ on the roof of its mouth. This is called the *vomeronasal organ,* and it can detect the tiniest amounts of chemical scent that were picked up by the trunk.

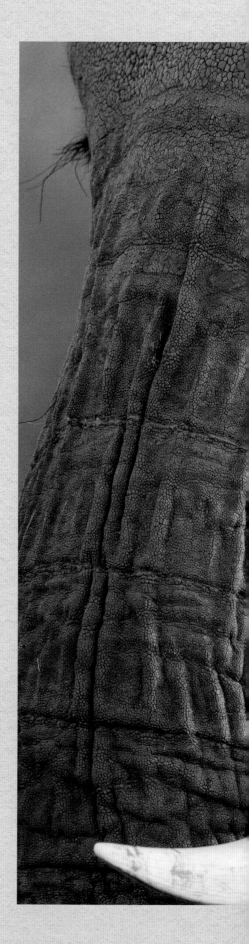

The temporal glands are located about halfway between an elephant's eye and ear on each side of the head. You can see a gland oozing a dark streak of scent on this elephant.

THE INCREDIBLE TRUNK

An elephant will use its trunk as a snorkel when swimming. Its trunk can throw sticks with good aim, or comfort a friend with a tender touch. The trunk tip has sensitive nerves like your fingers and can gently pick up a seed. Yet the trunk is strong enough to rip branches from trees. An elephant also uses its trunk to hurl dust on itself, to protect its body from sunburn and biting insects.

Controlling this multipurpose tool kit takes skill. A trunk has more than 100,000 muscle units, so it is tricky for a calf to figure out. At first, a newborn calf seems perplexed by this curious organ attached to its face. The trunk flops around and trips its owner up now and again. But by about three months, the calf will copy its mother by pulling up grasses. A calf will use its trunk to take food from its mother's mouth. It likes to check out what its mother is eating, and this helps the calf learn what foods are good. A calf will also pick up some of its mother's dung to eat. This may sound gross, but the dung helps the calf get the bacteria it will need to digest grass as food.

Elephants are good swimmers. An elephant can use its trunk as a snorkel.

African elephants have two points at the tips of their trunks. These are like fingers and are used to grab small things. Asian elephants have just one tip but are still very good at picking things up.

Learning to drink water with a trunk takes longer.

An adult elephant drinks by sucking up as much as sixteen pints of water in its trunk and then squirting it back into its mouth. A calf takes about six to eight months to manage this skill. Then it discovers another use for its trunk: It

This elephant has waded into the river to both cool off and take a drink.

makes a good squirt gun to spray itself. This is important as a way for elephants to keep themselves cool, because they do not have sweat glands. Elephants also keep cool by flapping their ears. This makes a breeze that cools the blood flowing through their ears.

TUSK POWER

Tusks first appear when a calf is about two years old. The tusks are incisor teeth, like your front teeth, only they keep growing each year. They can end up seven feet long on an old male African savanna elephant. Tusks are used for defense and display. They are also used to strip bark from trees, which elephants eat, as well as to dig edible roots from the ground.

During the dry season when food is scarce, an elephant will dig up plants with its feet and eat them roots and all. The elephant will shake the dust and dirt from its food, but even so, its diet will slowly grind down its teeth.

Elephants do their chewing with big molars that can weigh more than eight pounds each. Eating leaves, bark, and roots is very wearing on teeth. But once one set of molars has worn down, it is replaced by another set. There are six sets altogether, and an elephant starts using the last set when it is about forty years old. As those molars wear down, the elephant will have more and more trouble chewing its food.

ELEPHANT INTELLIGENCE

An elephant has the largest brain of any land animal. In some ways, it is like our own brain. For example, as a calf grows, so does its brain. That is because, like a human infant, a calf keeps learning new things as it develops. In time, its brain about doubles in size. Elephants can solve problems, and they have long memories. They are thought to have an intelligence similar to dolphins.

People who have raised orphan elephants say they have strong emotions. They can be happy, sad, playful, and throw temper tantrums when they do not get what they want. As they grow up, they develop distinct personalities, just like people do. Some are quiet and careful. Others are extroverted and daring. Some elephants can grow up to be bad-tempered, while others are gentle. A few grow up to be rather dramatic. Elephants are also very sensitive to the emotions of others. They will help their sick and comfort other elephants who are upset by petting them and making soft, cooing rumbling calls.

Calves are always learning new things. This one has just discovered how to use its trunk to blow dust on itself.

Elephants have sensors in their feet called Pacinian corpuscles that can probably detect infrasound vibrations traveling through the ground.

A large part of an elephant's brainpower is used to control its complicated trunk and to make full use of its powerful sense of smell and ability to communicate.

Scientists have now discovered that elephants even communicate with a "secret" language. As well as the calls we can hear, elephants make rumbling sounds that have such low frequencies our ears cannot detect them, although we might be able to feel the vibrations with our bodies. These sounds are called infrasound, and they are a great way for elephants to stay in touch. That is because they travel long distances. In the evenings and mornings, when the air is calm, elephants may be able to hear infrasound calls from as far as six miles away. At a mile or so distance, they can recognize which individual is calling.

Infrasound travels through the ground, too, and elephants can detect it with their feet. Fat pads on the feet gather the sounds, which may travel through the bones to the ears or get picked up by special vibration sensors in the feet. So elephants can probably feel sounds with their feet as well as hear them.

Elephants have such special senses that it's hard for us to imagine what the world looks like to them. It is certainly very different from how we see things. An elephant's world is illuminated by smell and sound. As it follows a trail, an elephant will swing its trunk back and forth to sniff scents on the ground. It will smell patches of elephant urine and dung and know who has passed that way, which way they were going, and even what they have been eating. At the same time, an elephant will

When traveling, elephants stay alert to their surroundings, relying on their ears and trunks to check for danger. These elephants have just crossed a desert landscape in Kenya. They are moving fast in search of water.

use infrasound calls to stay in touch with dozens of other individuals. In this way, an elephant constantly updates a huge social network that spreads for miles.

At even greater distances, elephants can probably detect the ground vibrations of a stampeding herd and know that it means danger. Even more remarkable, African savanna elephants are thought to detect the rumbling infrasound created by thunderstorms 100 miles away and use that to guide their migrations to new feeding grounds just as the first rains arrive.

A calf is drinking milk from its mother. Behind them, the wind is picking up plumes of dust, and beyond that is the volcanic peak of Mount Kilimanjaro in Tanzania, Africa.

THE AMAZING LIFE OF AN ELEPHANT

As an elephant calf travels with its family, it's learning all the things it will need to know as an adult.

By the age of five, it will be feeding itself and depends much less on its mother's milk. A fully grown elephant can eat 300 pounds or more of plant matter a day. It feeds well into the night. In fact, adult elephants need little sleep. Just two hours a night can be

enough, and they often sleep standing up. An elephant will lie down every few nights for a deeper sleep. If, however, its herd needs to keep moving at night to avoid predators, an elephant will go without any sleep for a couple of days.

A female calf matures when it's about ten years old. She will stay with her herd and become an allomother. By helping to look after her younger brothers and sisters, she learns to be a mother. African savanna elephants and Asian elephants usually have their first calf when they are about fifteen. African forest elephants have their first calf in their early twenties.

As a male calf grows up, he becomes increasingly troublesome to the rest of the herd. He may chase zebras at the water hole and generally misbehave. The females will scold him and let him know his behavior is not appreciated. Slowly, he will start to spend more time away from the herd, and at about fourteen years of age he will leave. He will go to areas where other adult males, called bulls, gather to feed. The young male will wrestle with the older bulls to test his strength. He may tag along with a bachelor herd led by a senior bull. The young males follow the older elephant like apprentices, watching and learning from their leader.

Young male elephants engage in trials of strength. This is practice for when they are mature and have to compete with other bulls for females.

As a male matures, he will enter a more feisty and aggressive state called musth for several weeks each year. This is when his hormones make him want to find a female to mate with. At first, he will have trouble competing with older bulls who are bigger and stronger. However, a male elephant grows in weight throughout his life, and during his thirties, he might become powerful enough to compete and mate for the first time. By the time he is fifty, he will be one of the largest and most powerful of the bull elephants in his area.

This bull elephant, named Tim, was one of Africa's largest and most famous big tuskers. He was about fifty years old when he died and had tusks so huge that they could drag on the ground.

A female elephant does not grow very much during her adult life and may only be half the weight of a big male. But she does grow in wisdom. Year after year, she will learn from all the experiences and adventures of the herd. Eventually, she will become one of the most knowledgeable and senior members of her family and take on responsibilities for its safety. She may take over

this role from an older elephant that has become weak. Other times, a large herd may split in two, and this senior female will leave with her own group of family members. Either way, her wisdom as an older elephant will help her family, during both good times and bad. Her daughters will learn from her skills and pass them on through their calves to future generations.

An elephant herd is ordered and peaceful.
Compared with other animals who live in large
groups, elephants waste little energy squabbling.

My favorite animals are often the small ones. I like to photograph spiders, beetles, frogs, and lizards. So elephants were a new adventure for me.

You might think that elephants are easy to find. But that is not always the case. The Bornean elephant lives in the rain forests of Borneo. The vegetation is so thick that you can walk within ten yards of one and not see it. One way to search is by boat, traveling the forest-lined rivers until you see one on the bank. But even that is hard. Bornean elephants roam an area of over one hundred square miles. You never know where to start looking.

Luckily, I had a plan. I knew a scientist who had put a radio collar on a Bornean elephant. This transmitted the elephant's GPS location, via satellite, to the scientist in Europe. He then emailed the location to me in Borneo, and I tracked it with Google Earth on my iPad.

As we traveled downriver, the small dot of our boat's GPS location inched across my iPad screen, homing in on the elephant. We stopped briefly when my guide spotted an orangutan and her infant high in a tree. The rain forest teemed with life. Bright butterflies flew rainbow colors over the canopy, and giant hornbills with flame-orange beaks plucked fruit like candy from the branches.

The hours passed as we continued, twisting and turning at each bend in the river. I could see from my iPad that we should be getting close to the elephant. But I couldn't be sure of seeing it. The elephant only needed to wander a few yards from the riverbank and it would vanish. Nature photography takes a lot of patience and a bit of luck. Finally, we rounded one last bend, and there, to my great relief, was the elephant. It seemed magical, only better. There wasn't just one elephant, but a whole herd grazing peacefully on the bank!

My luckiest elephant encounter was in Africa. I visited Amboseli National Park in Kenya. It is on the slopes of Mount Kilimanjaro and one of the best places to see elephants. I had a very experienced Maasai guide who helped show me many wonderful things. He also told me stories of the park's most famous elephant, called Tim. At that time, Tim was one of the biggest elephants in all of Africa, with tusks so long they touched the ground. However, I was unlikely to see him. Tim was a free spirit who roamed far outside the park, even into neighboring Tanzania.

Then one day, my guide got word from a ranger that Tim was nearby. Eventually, we spotted Tim feeding on some trees in the distance. We didn't want to disturb him, so we tried again the next day. This time Tim was marching purposefully through open scrubland, leaving a cloud of dust in his wake. My guide knew Tim well and recognized that he was heading to a water hole. Better still, my guide was sure he knew which water hole.

We drove straight there and parked by some trees. I was nearing the end of my trip to Africa, so this would be my only chance to photograph Tim. We waited for about forty minutes, hardly daring to speak. Then, suddenly, there was a puff of dust above the nearby trees and Tim strode into view, just as my guide had said. It was a precious and awe-inspiring moment. I was face-to-face with one of the largest animals on our planet.

—Nic Bishop

INDEX

Entries in *italic* indicate photographs.

BIBLIOGRAPHY

Garstang, Michael. *Elephant Sense and Sensibility: Behavior and Cognition.* Academic Press, 2015.

Moss, Cynthia J., Harvey Croze, and Phyllis C. Lee, eds. *The Amboseli Elephants: A Long-Term Perspective on a Long-Lived Mammal.* The University of Chicago Press, 2011.

To learn about how Nic created this book, visit **nicbishop.com**.

Elephant and setting sun.